## Advanced Praise

Soft as shadow twisting away from the white hot clutches of repression and complacency, *Ion Love*, by Julia Gaytán , is an alchemical equation of Brown girl feminism, hood intellectualism, and lip biting desire. Each poem births a pathway to not just witnessing the patriarchy's ejection from feminine consciousness, but an opportunity to transform our fingers into the claw and metal of our foremothers agony and rip it into a thousand precious deaths. Gaytán transports us into the throws of Audre Lorde's erotic, an invitation to push past fear, away from the "safe" swaddle of colonial traditions, and into every spiritual crevice of corporeal want one has ever lusted towards - be it fight, pleasure, or voice. This poetry collection is the back arch of expression and sensation while rejecting every piece of bullshit society keeps trying to feed its daughters; savor it like you finally know your worth and enjoy yourself as Gaytán explores the beauty of self knowing and the sexy in boundaries.

—Vanessa Rochelle Lewis, author of *Reclaiming UGLY!: A Radically Joyful Guide to Unlearn Oppression and Uplift, Glorify, and Love Yourself*

Julia C. Gaytán's debut collection, *Ion Love*, we feel the heat of romance and bodies in motion. We feel the blazing colorful sunset with "the anticipation of the pink moon" in the recognizable Los Angeles landscape that folks born and raised in this city are familiar with. She unapologetically explores the desire to be loved and seen. Gaytán shares childhood memories and past loves with an understanding of where she's made mistakes, with a deep sense of self-awareness that is refreshing and relatable. Her poetry is expansive and eloquent like the Fibonacci sequence, like the galaxy. She weaves past, present, and future, throughout her poems. Reminding us of the rhythms and movement of our beings on this planet, a rhythm all her own that spirals and unravels time and time again.

—Jesenia Chavez, author of award-winning poetry collection, *This Poem Might Save You (me)* and the forthcoming *Pocha Prayers*

In her debut *Ion Love*, Julia Gaytán proves herself a modern-day Ada Lovelace by way of Los Angeles. In this neo-Romantic treatise, Gaytán blends mathematical and scientific concepts with Classical philosophy in search of a trembling, breathless and ecstatic love. If you've ever speculated how Shelley or Juana Inés de la Cruz might've written about global warming, religious extremism, police brutality, Spanish colonialism or ICE, then look no further. For Gaytán, the perfected whole isn't found on the other side of love's equation but the naked need of the current moment is the only catalyst for an idea whose time must come.

—David A. Romero, author of *The Enemy Sleeps* and *My Name Is Romero*, editor-in-chief of El Martillo Press

Traveling through melodic rhyme and purposeful free verse, Julia Gaytán's *Ion Love* takes the reader on a journey across and beneath sand. Her work ratchets out cultural LA poetics with grit and heaviness. Gaytán's work bridges the divide between art and math, formulating mortality, love, and language into a collection of poetry that represents not only herself, but her community. From the opening poems like "Imported Sand," though the titular "Ion Love," and beyond "Motherland," Julia Gaytán's voice ricochets off every page, in the best kind of way.

—Matthew Mejia, author of *theFEEL*, Publisher & CEO of Nervous Ghost Press

# Ion Love

# Ion Love

## poems by Julia C. Gaytán

RIOT OF ROSES
PUBLISHING HOUSE
SEJATNGA
UNCEDED TONGVA TERRITORY
SOUTH WHITTIER, CALIFORNIA

Published by Riot of Roses Publishing House
Ion Love: Poems

ISBN (paperback): 978-1-961717-50-3
ISBN (ebook): 978-1-961717-51-0
Library of Congress Control Number: 2026945098

First Edition, 2026



For bookings or interviews, contact: jgaytanu@gmail.com

Printed in the United States of America
Edited by Brenda Vaca
Cover Art by Hoa Phan
Layout Design by Consuelo de Verano

To those brimming with relentless
passion, who create the electricity
that binds humanity

## Contents

# Section I:
# Identifying the Participants

# I Love You Conditionally

My love is conditional,
you know that

I do not love you as the Bible verse
or movie scene
or the never experienced memory
of an idealized love

I love you as the wild child temporarily entranced
by the kaleidoscope
who might, or might not,
remember that there is a world
beyond the scope

I love you
...as the gazania that opens during the day
but quickly shuts its pedals at the evening's start
...as the plant that is rejected by the soil
...as the earth that turns away from the sun
...as the awaited breath
...as the forecasted nipples that perk at the chill
or the anticipated soreness after walking uphill for miles

My love is not fundamental, as the laws of the universe,
not like energy, attraction or conservation

My love is but a composition that is temporal
and mutates through time

And yes, my love is time-bound
because love is not stagnate
love is not still,
but it is a love, still

# Ion Love

This love is loosely elemental like a molecule
It is born through the ups and downs of quarks, metaphoric quirks,
the push and pulls of the currents, literal circumstances,
the ins and out of hydrophobic miniature matter passing membranes;
remembering always that I flow, you flow, we flow...we are the flow

It is a love that is,
simply: it is

I love you as a lapse of time,
as a nagging thought that won't leave the mind ...
though it can, and it might, after time

This love endures the moans and groans
of day-to-day annoyances
It can endure the pauses birthed
by hurtful moments

If my livelihood is endangered,
it is a love that has the choice to abort

In spite of want
In spite of love

I love you because I do
I love you because that's where I am
I am in love with you

But it is not unconditional

Constantly vibrating
Emptying our voids
Becoming and being

We will grow, branch out, twist and shrivel
It is what is real
..this love is real

And though my love is conditional,
it is love just the same

it is love

## Imported Sand

Even our California shores
are for show
On the coast of wine and Hollywood, we import
Sand
of a sort
Art
or the like
And people talk too much without saying much at all
Not listen to our sea enough
Not see our sky enough

Don't wait for their turn to speak

Only appreciate art that has potential for
monetary appreciation,
depreciation of talents without price tags,
marketed name-brand rags
over the shoulders of
glamorized anorexic models on mags,
purple-tinted cabernet lips,
commercialized hips,
fattened cows, botoxed brows,
everyone medicated now,
romanticized vacillating moods,
overpriced
Whole
Foods
chain,
pushing undefinable "organic" grain

the bane movement of
"no GMO" rants;
although agriculture itself is genetically modified ...

but, what do I know?

I am renting
popular misconstrued publicized views of
singular truths,
though LA is multilayered;
multi-truthed

All I know is that I am also Los Angeles made
by way of Northwest Mexico,
the colonization of indigenous lands by raiding crusaders,
engulfed by seafaring landings, cultures transplanted,
influx of foreign seeds, the emanation of ideologies,
a struggle to retain a sense of monolithic communal identity,
overtaking and conquered lands, all part of my genealogy,
Arab-led muslim forces conquered Spain decades prior,
campaigned across the Iberian Peninsula,
both territorial expansions cruzaron al norte,
as so many did and do from military order or necessity

...from tortillas cooked on aluminum lined stovetops

From immigrants scared of being deported,
even though Braceros like my grandfather were legally imported,
the US GDP plumply supported,
though that fact is underreported,
just like the land that we're on being stolen and sorted

...generic cereal brands,
plastic bowls from the 99 cent store,
cleaning walls with "jax"
and 2-liter bottles of RC Cola in a fridge
that is never cold enough

From t-shaped poles that suspend tight wires, taking a
chance that our over-worn clothes won't
get swiped as they sway while they hang to dry

Always taking a chance...

like during our gambling nights of Conquian,
where every pull made a difference

like when a gun was pulled on my dad and cousin
while they stood at the door before we headed to my Tia's house
I waited in the backseat of the car, as I tended to do,
and knew to lay down when I saw the masked man with the raised gun
rounding the corner of our building,

as I smelled the dirt on the seat I thought about
my dad and I holding hands on our morning walks
to Jay's liquor store to
get La Opinión and Chick-o-Sticks;
feeling scared and safe

I was laying down for so long,
that I forgot to get up, minutes later I went inside
but no one noticed that I was gone,

... like dragging my feet to Sunday mass, begging my mom
to let me put the dollar inside the ofrenda basket,
Padre Nuestros and holy water
by the entrance door, praying that
my eternal soul would not end up in hell

Remnants of generational trauma, scarred spirit,
left from the rape of my great-grandmother
by a traveling railroad worker

...Kahlua mixed with café

from never attending overnight camps, shopping at thrift stores,
second-hand clothes, being on food stamps, and surviving on...

bolillos filled with carnitas
mixed with cebolla and cilantro to make it last,
Christmas gifts from St. Lawrence,
canned goods in plastic bags
handed off a truck on the corner of 112th Street and
Central Ave.

...pan con mantequilla y azúcar,
and beef flavored Top Ramen o Maruchan

from cigarettes inside ears to treat ear infections,
to melted butter to heal wounds

All these old and new roads traveled and crossed, quiet fears,
unspoken truths and a deep longing
to stay connected with
the home I carry inside

Because out there, we are treated as byproducts
traveling 'cross ports,
desired in some markets,
but undesired where it matters most...

all these truths residing in this veiled LA
the West Coast, Southern California, Los Angeles County
masquerade ball,

a tragic spiral yet beautiful façade

## Gelato and the Witch

I'll have to wait on the coffee gelato with chocolate chips,
it's unavailable, so I settle for the shot of espresso
over hazelnut Italian ice cream

As I agree to this concoction,
proposed by the young gentleman across the display,
you playfully bump into me from behind

I know you
and
I know that game

...it's one of my favorites

You wear your burgundy college sweatshirt,
the one with the fancy stitching,
and it makes me wonder how you were as an undergraduate
and if you and I would have "kicked it"
as we made our way through higher learning,
institutionalized schooling

I also think of how much softer your face
might have looked 15 years ago,
before the stress of mundane adult responsibilities

How young we must look right now,
gazing over treats behind the glass
like a small child being given the choice to
'only choose one'

I've always been bad at choices like that,
but I know you don't mind

You want a meal, I just want dessert,
and yes, I remember that you're not one for treats

I thought it rare, for someone as you,
to not enjoy the sweetness of sugar in food form,
since I know you enjoy its presence in prose

This affogato,
as you,
is an unexpected surprise;
the tasty combination of sweet and bitter,
brown and black,
cold and hot, here and there,
it is what I always wanted

I love it, but I glance over to your plate
and your sandwich, looks mighty appealing in its triangular glory–

I want it too

You came a long way to get that
pesto provolone turkey sandwich,
and I want a taste

You see the hungry look in my eye
and offer me a bite,
I knew you would, and so I messily rip
a piece from the whole
and eat but a fraction–
interesting pairing,
not quite what I expected,
but good just the same

It falls apart

The tomato makes an escape through the side door,
but I capture it and throw it within
the dungeon of my ravenous mouth,
because as a 'Sandwich Witch that has used her charm,'
it seems unfair to let that magic in the form of
healing herbs, cured meat and bread,
go to waste

## Aphrodite

A stone temple in dedication to your desire

Pause in front of the altar

My altered state is beautiful,
as yours

Your railway-shaking prayers
disrupt the solemn amber flames
that decorate this temporary temple,

flickering embers dance to the dirt below

I was formed by hands
in search of an ideal form,
but their lives passed
while I remain intact,
a standing form over decaying earth

Sadly, that's all I've ever been

While the cycle of life continues,
I continue,
as life continues after death,
perceptions of beauty continue,
as life deplores rudimentary ideals of...

beauty being tangible

Here I am, the goddess of love,
emerging century after century as a different form,
because of your spirit torn

stemming from self-deflowering, accentuating thorns...

Will you always fall to your knees in front of
my sculpted face?
Though now intruded by cracked stone

The dirt is soiled by tears gushing from closed eyes,
ironically leading to the *Wrinkle(s) in Time* that
crow the sight

Standards of beauty can be limiting

A place where individuals place their incense
in hopes of something more,
in search of superiority

And now this ideal is restricted
within white walls, behind the halls
that carry New Age hope
for aging people

Beyond the air and into the flesh

In front of the pupils
of a left-handed scalpel
carried by the right-handed wizard
of contemporary cures to
perceptions of the self

When will human beings realize that
there's more than this altar
created by destined frailty, more to this
life than dust reversed?

It is all merely brain waves
which, just as water waves wavering,
are tranquil after ephemeral storms

## We're Only Words

We write
That's our mode
Poetry is often the median
Free verse is the mean
Electronic correspondence is the plane

If we were to graph the points of our written work
We would be left with a bell curve
The hump coming from sensual exchanges; inputs
to functions having to do with desire
The output always positive--a modulus no doubt
The f(x) staying within the first and second quadrant
The first and second gear
The first and second base
Then, out of nowhere,
deriving even oddities,
having holes in our continuous stream of words,
creating non-existing limits

The first and second step-function
Acknowledging the gaps between
Waiting to plot the points we exhibit

I'm ironically cautious but reckless
Innocently flirtatious
Clearly elusive
Knowingly coy

If you take the derivative of velocity
you'll catch up to speed,
accelerate unto me
and see if we'll crash

## My Kindred Soul

in green pastures of my past,
you reveal yourself through
these frequent dreams of you
where we find ourselves in love
walking step by step
in familiar places,
in others' homes

naturally entangled
like the wind that carries away
the fallen leaf
rising and falling,
flowing with the tantrums of the wind

you guide me and I,
happily,
accompany you

at my feet lies the shadow of your being,
my hand enveloped by the firmness of your grip,
your rigid gaze forwardly fixed then there's a delicate switch,
softening with tenderness as you stare back unto me

that's how you have me at your side,
at every step,
all the while taking me hand in hand

there is no doubt that you love me, and I love you

we love one another as old friends,
known stories, kept secrets,
shared childish pranks,
accepting each other as we come and are

accepting that we will always find each other,
and by your side I want and must be

the world turns, and we remain,
in different guises, but at the end of the day,
with the same essence

you and I are inevitable,
like the sun that warms the river's bed–
waters that soothe and calm the flesh

the body that I meet in my dreams
is recognizable from this conscious life, but it is not you,
I wonder why your soul decides to tie itself to
a momentary recollection of his physical form

but it is your soul, not that body,
that calls me night after night,

the many lives I have lived are irrelevant,
just as irrelevant as the many lives
where love has not been felt
on awakening

I have lived your love so deeply
I have felt your love for centuries,

with every reanimation of my soul
body after body
temporary places
transformed faces
love after love

you find me in any medium,
in subconscious spirit
in singular universe
in temporary skin
in transient dreams
your love I carry in every cell

even at times when the sun escapes,
shyly behind the clouds full of potential tears,
you are hiding there

in this life our bodies are distant,
like sands separated by the sea,
but in my dreams
we will always meet
happy once again
living our temporary world

in these dreams I expose a half-truth:
the thread of time is stitched to the tapestry of passing sands
that know no better than to hold on to the blowing wind

## Mi Alma Gemela

en casa ajena, en los pastos verdes de mi pasado
te revelas a través de
estos sueños frecuente de ti
donde los encontramos enamorados
caminando paso a paso
en lugares conocidos:

naturalmente enredados
como el viento que se lleva la hoja caída
subiendo y bajando,
fluyendo con los berrinches del viento

me guías y yo, felizmente, te sigo

la sombra de tu ser a mis pies,
la firmeza de tu agarre envuelve mi mano,
tu mirad rígida así adelante
pero con delicadeza y ternura
cuando miras atrás hacia mi

asi me tienes
a cada paso,
me llevas mano a mano

no hay duda que me amas, y yo te amo

los amamos como amigos de niñez,
travesuras compartidas,
historias conocidas
y nos aceptamos tal y como somos

acepto que siempre estaremos juntos,
y a tu lado quiero y debo estar

el mundo gira, y nosotros permanecemos
con diferentes disfraces, pero al fin del dia, con el mismo relleno

tu y yo somos inevitables,
como el sol que calienta el agua del río–
aguas que calman la piel

el cuerpo que veo en mis sueños lo reconozco,
te he visto en esta vida consciente, pero tu no eres,
me pregunto el porque tu alma decidió atarse a
la memoria de ese cuerpo

pero es tu alma,
que me llama noche por noche,

no importa cuántas vidas he vivido,
o por cuántas vidas el amor no se a sentido
durante el despertad,

tu amor lo he vivido tan profundamente que lo llevo con mi ser

he sentido tu amor por siglos,

con cada reanimación de mi alma
cuerpo después de cuerpo
sitios temporales
rostros transformados
amores tras amores

me encuentras en cualquier medio,
en espíritu subconsciente
en universo singular
en cuerpo temporal
tu amor lo llevo en cada célula

hasta en tiempos que el sol se escapa, y, por
timidez, se esconde detrás de la nubes llenas
de llanto potencial,
ahi estas

en esta vida nuestros cuerpos son distantes,
como arenas separadas por mar,
pero en mi sueños
siempre nos encontraremos

contentos una vez más
y viviendo nuestro mundo temporal

en estos sueños expongo una media verdad:
el hilo del tiempo está cosido al tapiz de arenas pasajeras

## Blue Fever Sunset

A lighthouse stands perpendicular to the ocean floor:
its light swinging right to left like a pendulum–
seemingly parallel to the sinusoidal waves below.

The sunset is a grapefruit pigment with
a dash of Italian blood orange,
a blooming gaillardia
exposing its limonite core–
ore ripened by the anticipation of the pink moon.

The crickets' chirping verses rhythmically tread the path of pale spring
ground phlox
–vibrating to the melody of sparkling silver currents–
the underwater fields mirror garter snakes slithering through Idaho high
grass terrain;
both creating glittering slivers that absorb the sun's rays.

All the while I stand here amazed at the simplicity of it all.

# Section II: Determining the transfer for Electrostatic Attraction

## Bounded

Scalar fields are not bound by directional flow
but I am

Magnetic fields follow currents of influence
and I am impressionable
like melted taffy on a hot metal park bench during a
sunny summer day

Bendable like elastic
moldable like wet clay

Tug at me
until I am stretched between gritted teeth
and pinching fingers

grasping tightly
at the sweetness found amidst bitter struggle

## Ion love

My take on love
is the "I" on love

an electrically charged atom

an outer electron cloud
being ripped apart by the other–
attempting to find stability for its
unstable outer callers

rearranging creation,
one phantom
at
a
time

This union caresses
a need for completion,
or a need for disposal,
that cremates prior
rendition through a mutual
proposal

...getting drunk in subatomic love for 8
electric dollars
is a blessing

Hydroxyl presence
charting new synaptic saturated carbon paths,
my neurons find it hard to interact
because of the transaction between
hydrogen atoms,
exclaiming the point beside a
displaced covalent sentence,
a metaphoric transgression
in this
nucleosynthetic
venue

Peruse the elements menu
to describe my I on love:

sodium
chlorine
potassium
calcium
oxygen

The choice is limited by physical discourse
that allows a required connection to occur

I am an unbalanced outer electron cloud
making sense of the worlds I create
by an allowed circumstantial attraction

Like the one of lovers, future parents,
meeting through mutual exchanges of
spatial arrangements:
educational institutions or workplace situations,
that after a while of private interrelation
allow for an exchange of genetic information

Disclose these findings made during discovery

I object, irrelevancy!

Overturn the decision to allow
that evidence into question!

Digression

I wish to be the poet that gives the
romantic description of love
but I'll settle for being its paralegal

Regal pursue that leads to legal conclusions–
electron seclusion that leads to delusions

## Ion Love

Insecure of a connected certainty
like alkaline tracing an old river's bend,
so high at one point but now sinks
into its own depths--
beaming from aquatic eyes turned inward,
away from the sight of a depleting ozone layer–
radiating its awkward ego form,
deformed
by global warming across its waving
apologetic skies

demise
but
my id will suffice

the up and down quarks of protons and neutrons align,

the design of some foundational blocks
are perfectly cubed
to seclude strays--

betray this poem's root to the third degree,
dismay and gently lay my hexagonal fears to rest
over this frozen river's bed

interjecting into my orbital space
to create the initial interaction
that is sustained
through the
need of one another to fulfill
an empty space

faithful, deceitful or overburdened
electrons will bind
to my atomic mind

These loose particles will
peruse, complement and electrically bind,
becoming one with
my elemental kind,
and all I will become is an ion love

by an intrinsic shove,
to be completed outwardly

seeking balance through a needed part of another,
a naturally occurring formality,
that allows peace in a frantic subatomic reality

## Imbibe You

Te quiero tomar *like a foaming beer—*
poco a poquito, *as to not spill*
*the brown pale ale converted to white fizz*
no quisiera malgastar a *potential "buzz"*

Te deseo al amanecer—lo
fundado en la desesperación
de no saber el destino después
del trago... al rato contaré un dicho

O una mentira cruel sobre tragos
mientras discutimos los de estos cargos

Lágrimas ocultan *pitiless leer*
Me tienes tal y como soy; *distill*
*my habit of drowning in your sweet kiss*
*as you tantalize in my hypocrisies' puss*

Rato pensando, horas hablando
del gran espacio y la relación
entre humanos -- ¡Un último trago! Pues,
¿le entras? Ahora, mi amor, tu dicho...

Mejor tomar el tiempo perdido trago
por trago que desperdiciar lo amargo

## What's in a Name?

Rolling my tongue to enunciate your R's,
my mouth always left slightly open in the shape of an "O"
at the end of pronouncing your mother-given name

"What's in a name?"
But a history transfixed by imposed
consonants and accented hues
Assimilated frameworks
Sacred myths
Exaggerated truths

Mandatory European Spanish spoken then

Though now, a mere 5 centuries later
I thoroughly enjoy hearing you refer
to my *ma* as *"señora"* and address her with *"usted"*

It makes me nostalgic,
makes me think of the respect
expected at home—

makes me think of my *'apá*
*... que dios lo cuide*

Reminds me of eating *pan dulce* and *leche*
before bedtime,
after my mom's 8pm *telenovela* on Univision

I would sway side to side during commercials
and my brothers and sister would call me
*traviesa* and *coqueta*

just as you do now, with a wide smile and sharp chin

Your eyebrows as furry as my father's
Your perspective as keen as my mother's
Your name as "foreign" sounding as my own

You
own
it

It's the underlining curvature of culture

You understand the importance of having your woman's back,
reppin' your girl out in the world,
providing for the heart and home;
being a strong man,
and, yet, softening to my playful whims

You grasp the nape of my neck as I ride on horseback
like the Spanish *conquistadores* that crushed ancient temples,
usurped thrones, and plundered native lands

Plans gone astray,
we are the gray:
the indigenous and the white Spaniard
mixed; the mestizo

We are the dark brown:
the enslaved African, European colonizer
and subjugated Mesoamerican merged; the mulatto

We are the forced encounters, the resilient, the proud

We rise
like Huitzilopochtli

After I come, you come
like erected Aztec temples
that fell, we toss and turn what was there
and now I am below

like our forgotten tongues
buried under the rubble,
tragically covered by beautiful Catholic missions

After missionary love making
you bury your face into my hair;
tu cabello *castaño* prickles my pale freckled cheek—
like stars illuminating Tezcatlipoca's nocturnal sky
and they temporarily become a part of my *rostro*

Perhaps that's what's in a name:
the continuity of living,
giving and taking,
merging and pushing away,
warring and loving,
staying and leaving,
coming and
releasing—

always acknowledging that it is
the circumstantial prickles
that enhance one's sky

## I Want You

The motion of your tongue has created its own
neural synaptic path down my lateral cutaneous nerve;
contracting my thighs at the very thought of your
plump mouth dancing on the floor of my plumeria pedal lips.

Your wet kisses have left an imprint on my neuroplasticity.

The plastered ceiling in your bedroom can be so unkind,
when the splatter paint doubles as ink blocks that expose my
innermost kink, insanely wondering when the next invitation
to explore our humanity will arise.

Always a late-night lesson, a sundown surprise.

I wait for our next layover in the shadows,
though I should probably be in the natural light to dry.

Are we destined to rot in our own over-indulgence
from unmediated passion whose shadow grows long
as the sunny morning advances to noon?

Our bodies, naked, find comfort in the coarse sheets of your bed,
that mutually embrace slow, sleepless nights,
and the glaring longing of detached days where this frame cradles the weight of
another, our bodies are aligning to their natural design.

Touching our way out and into our photosynthetic
chlorophyll creations, relations that are one shade away from earth.

## Coffee

Let me sit in my own bitterness
Let the dissipated grinds flavor my ceramic taza
Let the heaviness of the bold embody me

...milk flows inside of me now
and the swirl converts the dark to a caramel brown,
the white crystalline sugar dissolves,
this composition becomes sweet enough to be sipped slowly,
enjoyed thoroughly,

Your awaiting lips delicately caress my chipped edges–
becoming frothed, while I'm glossed by the interchange

And after you're done
I am left stained,
tainted, a light brown color,
brown as the soil
where coffee plants grow,

but not
as brown as the coffee bean itself

## Just a Kiss

I beg, just a kiss
to shatter the silence—

a twirling tongue to invite expectation;
synchronized lips
to synchronized hips,
shadow our bodies swaying
along
the rhythm of each other's
near misses

"We're at our best when it's from our hips..." [1]

while our eyes lock; you proceed with a sly,
gentle, friendly grin
and I respond with a bit lip

Our chest cavities rise and fall,
breathing in and out, sweaty palms gripped,
fingers intertwined as we get lost
in the music from the snare drums
of our coinciding beating hearts
that resonate throughout our bodies'
Romanesque amphitheater walls

All the while our love is covered by
the traditional red, heavy polyester drapes
that hide the stage
but open
at the tug of
frail heartstrings
that hang
stage left

---

1 Cursive. "From the Hips." *The Ugly Organ*. Saddle Creek Records, 2003.

## Agave americana

The agave sap from your honey brown iris
traps my logic to a null state

I drink the mezcal that your searching eyes distill;
26 years of fermentation reflected by the taste of your
yielding pineapple lips digging through my receiving flesh
as the pinch distracts from the tracks of your sugar flow

Each crevice of my blood stream
caught off guard by your harsh,
slow-running course

I must blow
out
after a shot of your eyes,
so the taste does not overstay its stay
in the mouth of my oval lens

Oh dear, I know how our drunken raucous nights
hide the subtext to our locked glare,
though the intent is concealed
by our whispered sensual verses–
the loci of universes

The properties of our attained identities
go unmentioned,
and are compromised, redefined by the performed exchanges
that fit the genetic mapping of our sought
humanity

a fragile reality

this psychological figment to the imagination is
such a striking refracted visualization!

Soft kisses under the mediated moonlight--

the beam hazily peaks
through the heavy Náutica rayon/cotton drapes
that hang along the chipped wooden windowsill--

providing a muffled visual
to the entrapment of sound waves
that expand
across the vacuum space
of your living room

The echoing waves rifle
through the ambiance that surrounds
as I recite poetry and, at times, strum my guitar
to the b-side of tapes

fueling the vehicles of sound
that crash against the curves of your face

forcing air molecules to rise and fall at a periodic rate

Fate is coordinated
by points across this interrupted plane

This still space dances
along a crazed dark state of familiarity
where your touch erases time

The caves of your pupils elongate the path through
the yellow brick road of green differentiation,
the discoloration of complementation,
a reversal of photosynthetic progression

But I stop myself from getting lost in prose so I start

retracing tracks,

blindly walking down a lightless path that has no end
in sight

and
regrettably

there exist blueprints to our allowed existence,

whose interpretation is relative to our being's status

Our place in space
Our movement in time
Our self defined

our state of seeing...
and being seen

My natural ocular state is -.425 on each eye
(with a minor stigmatism to the left)

It's unfortunate that I ride life
through opaque blurred lengthways

phases of sinister steps,
sinistra routes,

the venture point of venturing through Earth's
transcendent curvatures that
fine tune the culprits of a sought, yet impossible,
left ocular dominance

I am a cornea,

an unchanging module forced to remain
in its pictorial state along the path of this
stipulated function.

And though
I can register the delicious agave americana
that finds its flipped image moving along the
ordinance of movement

I am no fool
and I know
it is a fleeting image

that time will blur

## Honey

Honey as dark and rich
as the deepest oil well
drips out

As sweet as the juice of sweating berries
on the face of vices

Dripping

Covering virtues
Manipulating façades

Jading the principality of honey

The pleasure that coincides with honey and me
is not that of the absence of pain
or the absence of anxiety
or the absence of uproar;

it is quite the opposite of Epicurean thought

because its sugar makes me unstill,
makes me want more,
such that I will never have enough

I still have the fear of loss
that leads to my own fear of want

You are the honey that taunts my lips,
spreads over my toasted thighs,
melts unto my tempered flesh
and I need more

More!

Less is more?
No!
I don't believe so!

Your jar of honey is at constant demand
Like transported oil
Like soil
Like copper coil

But...
I still have the fear of loss
that leads to my own fear of want

You are the honey after its spoil
After its drip
After the destruction of what's underneath
After the loss

After I figured out that honey is not always
good for my health

You are the honey of distilled grapes–
beehives transformed to barrels
kept underground

And I want more

...I always want more

## Poets Make Selfish Lovers

Throbbing aching thighs that eagerly wrap around
the sensual world, as if it were a sugary sweet melon
within our open palms waiting to be ravished by our
arbitrary observations and rulings; eventually,
breeding the zealot seeds that inspire poetry

Minds that perceive through the aid of idolized human experience,
that reach into the mediocre for reflection,
from the metaphysical to the tangible world:
the conversation, the kiss, the view, the scene,
the food, our bodies nude
and, the feelings we try to conserve, ironically,
by expressing less in person to express more on paper

After making love all day and night
to temporary circumstance, we always want more

more than the moment, more than the glance,
more than the quiet trance,
more than the poem that finds itself below our hungry stare
more than the lover, more than the friend, more than the job,
more than our social stance

more than their body's climax

it is no wonder that we choose words to
unfold the quilts of normalcy
to define our feelings of longing to boredom,
trust to lust,
contentment to melancholy...
rarely being satisfied at the outcome

unsure if we've received and given what we truly need

## Act III: Scene 2

A mixture of lust, unacquainted love, and deceivable resolutions

The moon is in love
with the unseen clouds of vanity-
the other side of the turning scene

Longing time frames that search for the end result

For the drowned sorrows of hubris and oedipal complexes that kill
our passions through and through

Until we're left with nothing but...

but no,
this is not the case, the
moon knows not to love
but to
want and not acquire;

to tempt with the
mystery of virginity,
and listen
attentively as the wind gloats about

her relationship with the clouds

...the wind whispers of the rain's betrayal

The moon denies her love for the clouds,
(or mere lust for the limited known)

She has fallen silent
in the presence of the wind
for the purpose

of mere example

# Section III: Electrostatic Repulsion

## Maudlin Impressions

Regret has numbed
the passionate emotions
that my mind once allowed
my neurons to transmit.

Admitting only the disappointment of lack.

A pernicious knack for many
reduced to
the one.
I wanted the one
when my feelings were only
black or white
but now the binary has changed to
0s and 1s–
the countless combinations
have pushed me to override this
single-seeking over-driven
multi-drive
hard drive.

Monogamy has won
the competition for media space
in a growing polygamist
commercial global market.

I have sold out to the best offer
of one socially accepted offer.

I am digitally nostalgic for drunken blackouts
and shrunken forgotten numerical expeditions
across the software
of my algorithmic being.

I have been immortalized by
seductive transgressions of
congressional selective acts
that have popularized this network,
that offers no privacy.
2 additional uploads
have set security parameters to
circumvent my allowed relation.

I allowed your
advertisement firm
to take up all the media space
that was left to vie.

It longed to be auctioned at one point
but now
is simply there for the selling
of mere pre-set impressions
offered at half the value,
for fear of a minimizing demand.

Just as software once installed
has a "No Returns" policy,
I have a "Regrets Guarantee!"

And regret always sells.

It is the food of poets distraught by the exposure
of their wounded gigabytes.

## Your Radiant Saddened Grin Dulls an Already Somber Day

Contagious to the kiss is the sickness
of ease that spreads across
your smile-ridden lips;
an aphoristic truth
that spreads as a pathogen
disguised in a mournful complicated glow

And yet, painfully, your emotionless smile continues
to dull an already gray and somber day

I feel like an energy wave experiencing amplitude decrement,
taking less area, feeling less me

I grow stagnant as a ditch's bed

Your lethargic smile offers nothing
but an ill-accepted pacification that bounds me

I become ironically tense,
wound up by the listlessness of your lazy smirk--

The physical structure of it will remain,
temporarily tucked away in time and space,
fossilized as a carbon-based life-form decaying,
oxidizing and rotting the frown,
to then be refined as a passionless smile
to fuel the machine within

After years of rehearsing the same play
of socially accepted optimistic displays
of a happy character, a constructed and executed plot,
a climatic bittersweet end,

to then be left with your
default resting smile
which continues to make it okay

Though you're not okay
and I am
a melancholic alcoholic,
a tortured fool
in search of a potion to cure all ills

This pill is hard to swallow
so I wallow
in liquid form defeat,
fluid in my own non-crystalline depletion

And it cannot be denied that we are stuck in phase
With the same saddened grin day after day

Yearning to replace this story's end
with a passion
unrehearsed, undefined,
unaccounted, yet understood

For the day will continue to pass,
somber as it is,
but with the hope for sentiment felt
or a new fervent face
staring unto my
displaced
throbbing
ache

## foreign struggle

our strangled love grows as the tree held down by weeds
supplied by loose, untethered, scattered, seductive seeds

freely roaming Oriental Bittersweet, took root at the feet
secrets secrete when their nature is smothering defeat

a slow, dramatic, choking death for oblivious trees and bushes,
bird-transported death sentence, by wind or water pushes

invasive species decorate the outer bark with girdling vines
nails crawl along the outer wooden skin leaving scratching lines

bulging veins, pumping redirected earth
the upper reaches of organic structure rupture at rebirth

its trunk offering the road up to branches choked
by their reaching fingers that crawl, entice and provoke

untended fields, unnoticed, hushed and silent
the end is gradual, leisurely but violent

the tree once strong, with roots that held, embracing land,
is now too weak and fragile to hardly take a stand

but the red intrigued and nature took the course of play
held down but held as one, both withered away

## Settler-Free

The hunger of desires
form lust,
the outer crust
is renewed by the creamy white soil
of distilled water, slobber, crippled
by the thirst,
clobbered by the mediator
of the word
and the thought

I revel in the layers and clumps
of untilled fields,
caressed by the attentive wind,
drenched, occasionally, by sporadic rain,
heated by the commanding sun,
nurtured by the giving earth,
prepared for growth,

a land that birthed...

...you packed your bags and
became semi-sedentary

You walked away, at your own dismay,
from a land that produced
greenery and fruits that fed,
sustained, maintained, instead
you chose to denude the land,
to excuse the peruse, the wander and seclude,
to then, remain distant and at bay

like the endless line of men who stray,
as they pass the midpoint of the circle,
strolling from one half to the next

Joining an other, during this
tangential search and find

Of course, her legs are stronger than mine
'cause mine were merely meant to wrap
around your waist like the product of diameter and pi,
while hers were trained to run alongside then lay
with those that merrily traverse her inviting river's bed,
embarking on a new sector, on a new arc along the way...

Maybe you were never meant to stay,
Maybe you were never mine,
perhaps it was wrong to colonize
or to even set foot on unbuildable land
because there is no *terra firma*,
just a *firma* on a light peach-colored paper
that held the promise of permanence;
chemically embossed,
and watermarked

a fallacy, a myth, a lie,
a blithe fantasy

purely performative, two-dimensional,
taken for granted, diminished and disregarded
like land grant-Mercedes

I'm vying against the Euphrates

wishing,
even though
I have no coin
to throw inside this river's bed,
no magic lamp to rub,
no red dancing shoes to tap
no shooting star or tunnel
to hold my breath within,

all that is remaining,
after being left on these abandoned plains,
is utter denial, shock, and bargaining,
then, thankfully, the realization of:
it is better to be alone than ill-accompanied

finally at peace, left alone to grow naturally,
without the heaviness of constantly
tending to the rigors of pseudo-settler male necessities,

yes, tended land has its own allure,
like promises of forever that are officiated,
endorsed, boxed and sold,
but contracts, no matter how advantageous, are still ignored,
declared null and void, so I'll be

riding the gush and blow of solitude,
breathing slow, laying low
solidifying strength of the inner-know
as to exist in the spirit of the land and in the soul of the soil

Ion Love

## Obsidian

Pulsating alizarin bubbled
from the mouth of your
deadly vent
eventually erupting,

running rampant,

gushing
over
my long lush valleys--
mercilessly burning down the Red Firs
that gave livelihood to my being,
until finally settling into a river's bed

You would think that all that would be left is
volcanic ash,
dust and
uninhabitable land
but no,
obsidian was created after your
destructive toll--

the edges sharp enough to be used
as a lethal tool
yet malleable enough to be chiseled
into a shimmering raisin black stone–
smooth breathtaking prose
that will
complement the
fluid crimson ink
spilling
on this new blank sheet

Regardless of the devastation
you've left behind–
obsidian was created
since

I treat lovers as I do poems–

as dwelled upon interpretations
that become secondary
to the painful experiences that can be reworked,
reframed, replaced with...

a poem to be written by the swift strides of
the spewing pen–giving meaning to time spent

chipped away
like obsidian...

Felsic plasmic passions
froth at the surface of the tip
to only become placid
at the release of gasses–
expansion of syllables promoted to words,
chords of residual naturally selective
discourses;
designation of
sodium, potassium, oxygen tracing,
nitrogen racing, as I'm facing
the metaphoric cremation
of a life I knew before–

And though I was
predisposed to your vindictive form
I never really
expected
to be burned
since
I grew accustomed to the tranquil scene,
unconcerned with the rumbling below

I should have known,

but regardless,
I am still Mother Earth
painstakingly transformed:
punctured by the fusion of atoms,
merging to a new state,
forming molecules, freezing movement,
breaking symmetries,
and
the end result
is still
beautiful

I am still beautiful

## Jazz Induced Poetic Cacophony

It takes one amp to stop the heart–
it has been two
and now I'm about dead

Your absence left its rotation
Tethering the blanks
The probability changes with your every
appearance,
then disappearance,
then appearance,
then disappearance,
then ...

the music starts again
with the same predictable melody
but it's interrupted by an
inspired improvisation that leaves me at a
breathless awe and awaiting,
still without the comfort of knowing when you'll come
back to me

You are both the harmonious music compilation
that I replay over and over again
and the sporadic discordant cacophony
that I hate to love
but love nonetheless

## you and I

I need the difference of our coordinates to be oddly numbered,
to complement my even rate of change,
each second the secant line is redefined, approaching the point

where a tangent line forms

you and I,
following the ordinance to the function we encompass,
we will become infrequent points of reference

...a woman you once knew...
... a man I once loved...
as we rode down a pre-set course,
not even in the know of our predestined flow

the function of my being
is distraught by dependence,
independent of your distance...

even though the instantaneous rate of change is
dimly noted

## At Least I Got a Poem Out of You

I did not get the happy ending
or the love I long to feel for another
I did not get the hurt or confusion of wanting the other,
just a poem that I rarely share with peers

I only got the empty blank verse-
perverse perception, prescribed medication,
wine-induced numbness; the cowardly way out of potential pain;
not letting the broken promise resonate

One mute day alone, months thereafter,
filled up with the painful noise of neediness
to then selfishly lure you to be mine at my convenience...
I hope my *Bukowskian* narrative is

worth this poetic cacophony,
the ripened metaphoric presence of you
and the naive contractual agreement of separation
of our bodily estates when carefully

prophesying our lemon tart end,
salivating the sweetness of the first bite
as one compares the new to the old, typical resurrections
of salvaged food leave me unfull and hungry

for passions that are in black and white,
carefully chosen words; counted syllables
for a poem that deserves no other name than what was left here,
a gnawing for transmuted empty sentiment

## i can't do anything but hide my scars

i can't do anything but hide my scars,
dress my pain by wearing beautiful dresses--
the crisscross colorful patterns
divert the eyes of those who see me,
they focus on the fabric, and not my pensive eyes
they are hypnotized by the way the fabric sways
as i step right to left, right to left
my legs coming forward one after the next
i keep my voice low, flirtatious and alluring
so they can give in to my selfish directives
entice them to adhere to my discretionary needs

they think of me as an unseen constellation,
only visible through darkness,
they think that since i whisper to them
that i am a good listener
i am not
i have learned to play the game of longing
belonging, dislodging
i have learned to look into their eyes in the morning,
make them feel like they're more than a passerby, but
i say a quick goodbye before the birds can sing,
or the apertures between the blinds gleam

i pack my feelings in a knapsack and leave
back to the barracks

during that time i developed the skills of
provocation, disassociation, dehumanization,
deprivation...
i have learned to cover my tracks
but i always leave a trail along the road
i guess i'm a "soldier of love."
i want to be found,
kept safe for a while in some unknown haven,
or for the war to be over so i can go home

i just want to stop fighting in this useless war

i wake up, suit up, and report for duty, dutifully

curl my eyelashes so i won't curl unto my true self
paint my nails in bright purple and red pigments,
to distract them from what my hands have done
fragrant my skin with flower scents
to suppress this dreadful, dubious stench
i don't need anyone's expectations pressing,
gravity is the force that keeps me together,
dark energy between my atoms bind my consciousness:
like heat to the flesh,
like flesh to the soul,
like soul to the spirit,
like spirit back to the flesh,
and i feel so empty
i feel like the space between
the scar tissue and the wound

the internal fault lines have shifted,
and i have fragmented many times over
dispersed cracks;
the screen of my being shattered

yes, i feel like the shattered screen,

the clear plastic silt barely visible,
barely keeping me together

Ion Love

## If

If I wade in the wind of whispered words in wanton wilderness,
will the reverberance somehow fill, though not heal,
the hollow space pulsating inside my chest?

Will lust resonate loud enough to silence
a phantom heart left yearning?

Will the rush of being a muse, amuse the wistful gentle glide,
or leave a hunger for the heavy-handed stroke of the engorged brush,
blush too faint of a technique,

slap the blob of paint unto the other cheek,

cause a squint that lasts enough to excuse that my searching eyes are
no longer being reflected on the smudged mirror screen,
superimposed on that rickety closet door, that never could close

The wheels are off track so they never could hide the costumes inside

If I hold my breath so as not to upset the intrusive, demanding, blank walls,
will they start to realize that the vibrations coming from my voice are not
there to move air particles that eventually bounce off and reach ear canals?

If I get stuck on the hot asphalt of a newly paved road,
will our tiled floor notice that the weight of my mass
is no longer there tracing quiet steps over its loose discolored panels?

Will our home's troubled spirit ever accept that
what was understood to be permanent was just a
temporary lapse of reason,
that our monetary savings did not save us from
the poverty of self-depth...

Will the moon's crescent have no home to light on a starless night;
after its disappearance into the other side of the globe?

Will the sun's rays still noisily creep through our bedroom's flimsy white blinds
if my long curvy body is not there over our tousled sheets to receive
the tentacles of heat that have grown accustomed to settling their weary heads
onto my freckled tan skin?

If I go, will our home realize that our stars have not aligned
even after all the years of tracing constellations;
even then, always knowing in the depths of our being
that we were viewing different skies

## Sometimes

Sometimes the overflow of an ocean
can lead to submergent coastlines,
becoming the inland bays, then submarine canyons
of wrinkles of too often temporary frowns
turned permanent

Outlined by the salinity of emotions
that erode the cells of my youth

Muscles strained as they bluff their way
across the map of my face
segmented by the political lines of advantageous relation

That sometimes cut my hypocrisies

Sometimes because those lines could be moved
changed
arranged
gerrymandered

Then, after perversion has taken its route,
a set geographic reality becomes a distant memory of
"some time ago" before...

Where we realize nothing is constant
even love is temporary
and lust, the passion of the world, is seasonal
and easily applicable to anyone

Where anyone is the sometimes of every time
and their temporary permanence becomes a permanent present
that camouflages socially constructed temperance

Framed and hanged from a sideways wand,
fastened by the tight parallel of my needs and wants
and I'll wish my way towards something that does not rely
on luck or charm
or until time is up,
and fasting has gotten old and tired
and the next step of even more fulfillment is required
for survival

Something that sometimes will not go,
to then reappear with the flip of a glass,
the drop of a sand pebble determining my future
by rewriting the past by moving sands

## Impossible Encounter with the Other Side of Our Truth

Is the Moon the somber neighbor of the Sun?
Both tenants
of celestial galaxies that enforce
bright fierce stars
upon the stark living room of Creation

The Moon turns away from its light,
its subjective gaze redefined
by chosen sight, like a picture
Confined by the peripheries
of galaxies extended far
and wide... but narrowed by human's
supposed logical advances
in thought, resolution and "Right"

As that of... Justice, Equality and Truth--
Deceptive
Equilibriums! As that of Good and Bad
Bipolar
unilateral cognitions! Limited
retrievals
of complex ideals as that of Purpose

Limited by philosophers' corpus, theorists' focus...

Yet can we define the continual from
infinite?
A spatial Labyrinth can only extend
as far as
the mind can perceive, but yet still can exist

Without acknowledgment, notice
acceptance or recognition
Maybe, perhaps hidden, within
Interrelations of imposed
contextual definitions--

Human superimpositions
The connotative restrictions
of memory, attribution--

Without set mathematical proportions, value, or structure

Inexhaustible
Never ending
Limitless
Boundless
Vast

That is the internal way of finding the zero
How one portion of ourselves could blind and restrict
by the ignorance of possibility

of our Orbital Space

# Section IV: Strayed Electrons

## Fish

out of water
in air
the medium changes
but I am still

present

filling hungry gills

## I am the best type of shallow

My paperweight words hold no weight,
too cautious to be truthful, too thought-out to be real–
appearing to be complex and deep
but they're quite shallow
as I am

But I am the best type of shallow

The type that sees where you're at, and lets you roam free within me
The kind you can splash within without feeling the fear of drowning
The kind to learn in; the type of shallow that welcomes doggy paddles
and erratic strokes without judgment

The kind to wet your toes, knees and thighs,
but not your chest cavity–
never your heart,
never your head

Not with me,
because
in a way,
I am the worst type of shallow

The type that offers slippery footing on solid ground
The kind that finds satisfaction at your longing
The kind who carelessly and freely gives away wet kisses

A public pool of sorts

The kind that finds solitude at its waters staying still

The kind that gives you false confidence
and encourages you through calm waters to venture
into the deepest of me

Only to slip, swallow and gargle
this sanitized, over-chlorinated water

It rushes into your mouth, out of your nostrils,
finding its way into your lungs,
you are, again, splashing erratically but now it is to survive

Then, all is still, you are completely filled
with the regret of deciding to take a leisurely dive

## night belly

still fog hangs in the air like a loosely held noose
hanging comfortably from a blotched bloody finger,
injured by the strangle, glossed by the tears,
heavy only by circumstance, proudly but casually held

the ether salivates at the thought of waving particles moving
space, rumbling heard from beyond the murky lake,
the hungry belly of the night sky letting itself
be heard by those who, unwillingly, find themselves in the grips
of its wanting snare

the thick stench of daisies and regret permeates the air,
filling the forest with the desire to bury the dead, shift the scene and
ignore the heavy sobs of loss

the flicker from a clicked lighter, momentarily sparks,
and briefly illuminates his malicious smirk
his hunger for stillness only rivaled by the night's hunger for company
inside her estranged home

Ion Love

## Besides

If I were a poet
I would write of the other

I would write about the one beside the laugh,
the one who is hard to comprehend
outside of the uniparticular hyperspace
of private conversation;

an interchange within a microscopic cube
whose walls receive vibrations perplexed by
personal context:
symbiotic perturbations

paused,
waiting,
caught in spatial time-

until the responsive action of fleeting
temporary words flow;
feelings described, thoughts received,
ideas explained, though only understood
by the one
in the shared consciousness
of the other

But I am not a poet,
I do not have the classical training of expression

Besides,
I have more to do than this

I am a mere
brown-haired girl
sitting at a crêpe cafe,
writing rapid thoughts while
drinking sparkling California wine
before noon time

as people share themselves with the other,
I share myself with
myself

A liquid state incurred to cure boredom
motivated by the insecurity of
being the lone spectator:
a social being caught off guard,
in need of a
brief private interrelation

But I am stuck in the margins of the story,
I am not in the plot,
my place is in the setting–
an eclipsed extra

I sit here yearning
with the supernovic desire to be placed,
replaced, with the one beside the other;
to be the welcomed occupant in
the place setting on the square table
of human thoughts exchanged,
but instead
I emit displaced glances
thrown unto the unseen strings that void
Planck-sized unfoldings between the manifolds
of interpersonal circumstance

And though I am not a poet

I write

as to place myself in the coveted position of
the narrator of exchanges that occur outside of me

As I sit alone, beside myself

## Between Classes

Sentada beside the shrubs,
con libro medio usado y libreta nueva,
I attempt to theorize, poeticize...

While you water, dig and plant
I distract, attract, create the abstract
from tangible hands–
the noun of your presence analyzed

I write, "... hands like my father who made furniture,
both you and he like Senju..."
followed with, "...noticing you notice the leaves' dull green,
sharing the same plane of indigo skies outlined by
waves of green discoloration..."

...on the west the sun sets

trabajando on the west side I have come to understand the eclipse–
the world between poverty and over-indulgence

the current of these opposing forces crash against
the shore of my reality, this variation of truths offers a refreshing salpique

Tu cabello castaño cae sobre tu rostro, refleja la luz del sol,
reminding me of mudslides in Villa Nueva durante el invierno

your silhouette forms an uneven shadow on the backdrop
of bushes that I just trimmed

Sé que me ves, pero ¿me reconoces?
Porque, mirar no es ver, y ver no es entender
Without the water ¿quién soy? Si quieres saber, te digo

Soy el padre, el hermano, la tortuga y el conejo
Soy la sombra en el asfalto y el sol que quiebra nube blanca

El maestro y jardinero
La luna influencial y el árbol derribado
La mirada fija y el reflejo empanado

Estamos entre clases

Los dos venturing through dual-consciousness
and even though we're separados por edad, profesión, y educación
los dos caminamos con el espíritu de nuestros antepasados
a nuestro lado, destinos atados a estos mismos pasos de esta Universidad

## Humanity on Pause

Christian medieval art portrayed the need of tragic blind faith
through archaic juxtapositions of heaven and hell
to scare one into a dogmatic viewpoint that limited one's social evolution
in order to ensure complacent surrender of one's consciousness
in exchange for a competing group identity

I do not want to be absent of desire
I do not want to detach myself from want
I am not a Buddhist at heart,
though, I want to be at peace like one

Piecing my transparent colored pieces
to form a coherent scene

a medieval mosaic placed between
the outside plague-infected world and the
inner celestial promises of the church--

biblical promises of eternal peace
crumble at the thought
of, what want is not taught?
of, what *truth* is not questioned for truth?
of, what heaven is not contextualized
to fit the frame of material and spiritual things
not acquired by our eventual mortal
and, for some, moral decay?

Even now we let this want of heaven
blind us from who we are

The fear of hell or social critique
keep many of us silent from our desires
such that we're physically alive
but, ironically, spiritually dead

The gargoyles of gothic cathedrals
serve as reminders of the threat that awaits
if we give in to our human adaptability
of acceptability

Acceptance of change
Acceptance of struggle
Acceptance of premarital sensual desire
Acceptance of questioning the government's stake of our spiritual estate

"Acceptance of that outside the scripture
is seen as being lukewarm in the face of God,"
Brother Gabriel preached during Sunday mass

Revelation 3:16 reads:
*So then because thou art lukewarm,*
*and neither cold nor hot,*
*I will spue thee out of my mouth.*

Rejection again.

I am reminded of the God Virus by Darrel Ray,
who's still writing,
while Avijit Roy of the Virus of Faith
was hacked to death in Dhaka,
Raif Badawi imprisoned,
Salam Mema martyred

Attempts at permanently silencing critical discourse,
to force a course of supposed piety and humility,

Attempts to justify murder and the taking of land,
to force inhabitants on inhabited land

so many lives...

Would I be as brave as they are in Bangladesh? In Palestine?

Or so many other places where free-speech or region-born
can imprison, maim or kill?

I can leisurely write,
left alone to enjoy the freedoms of my circumstantial place,
laced by the electronic blindfolds that instantaneously entertain away
the pain of seeing so many lives being so unfairly discarded,
regarded as less...

I can choose the sequence, pattern and color of the fitted sheet
that finds itself between the receptors of the "facts" of unfairness
of another place, injustices that I do not face,
hence, I don't fully understand...

We can't be free with limits to expression, but who am I
but another American knowing only my own context?

Turning bloggers, journalists and authors imprisoned and murdered
into mere poetic subjects...
...such an unnatural state of advantageous relation

The disbelief of differing existence is our biggest sin,
our ignorance toward the "other's" circumstance only makes us
entitled to think we know more, that we know better,
that we know what will fix our neighbor

yet a disbelief of our shared humanity,
and influence, leads us to inaction,
thinking we can't do more

My own humanity placed on pause because of a privilege,
and ignorance,
I have yet to accept,
so all I am left with are words on a page

## Morning Dew Turned ICE

Quiet contemplation
interrupted by
the tap of a swaying tree branch
that
drops formalities at the shudder,

exposed
trunks are beaten to a pulp
but the paper never comes

As the California Poppy opens
the window sweats a cool drop of morning dew
over the chilled white windowsill, ice forms

Whoosh, drip, blow,
BLOW!
BLOW!
BLOW!

the shot too loud that eardrums burst
the bullet cuts, ripping clothes and flesh
the buzzing noise of a soundless ring

Nothing heard
only seen
through the dusty window screen

only felt by the memory
of sound

Sour visions
in a bitter world,
breaking the glass, shattering,
tearing the tear of the tear of
this timeless tea

Will the warmth of the day eventually melt
the ice that has shut the window closed?

## I'm Earth

I'm earth, do you see me?
With your snide rolling eyes,
unaware of my tides,
or my cold ocean stare,
my tight ozone sun-kissed lips
become slightly parted, unzipped,
ripped, surprised at the audacity of
your response
to my
body's
sickened state--

cutting me down to pieces,
the parts that give you life,
are replaced by open-farming livestock zones
subcontracting death in methane form,
causing a rise to the boiling disappointment
that is bubbling from my pours,
my ice cap brow melts and drips
from the trapped heat,
sweat gathers
at the land below your feet

I'm developing a fever at your touch and yet
you refuse to diagnose me

You are the cause of my symptoms

I wonder,
when you will cease to destroy my membrane walls?
When will your CFCs stop attacking my
CD4 white cloud cells?
The alarm bells ring, and you act with impunity,
aren't you alarmed at this weakened immunity?

I only want to provide a good home to you,
I had it all set up,
but instead

you appear to be a hungry virus--
taking over more and more,
and it's never enough

Your actions have unintended long-reaching consequences

-but you won't see it-

like
high-class patricians
unknowingly exposed to toxicity
by partaking in the heated glory of
breathtaking Roman baths,
transported by ancient aqueducts
whose pipes were laced by
poisonous lead,
only now it's Flint, Michigan
and the decision from Lake Huron to Flint River,
to save a sliver of silver,
was government-led

just like a moment's deafening blast brought upon
the push of a button thousands of miles away
--triangulated drones
dropping bombs,
the names of the casualties unnamed
renamed as a number,
the remaining survivors
maimed, land cremated,
to be reanimated at a
later time and date

You can dress it up in Lavender
but death still smells like death,
like iron, copper ... metal-like

no critical thought of how your actions affect humankind

Racing crimson crimes,
slaughter of fathers, mothers,
sons and daughters only seen as a
"needed cost,"
a game of catch and toss,
not empathizing with the heaviness of loss,
of a never-waking nightmare of unjustified demise,
then surprised at why some radicalize

All these technological advancements mask
that social intelligence is at its infancy,
the need for comfort, and ironic fear of the other,
surpass a common sense of survival

I'm earth, when will you grow up?

Sibling rivalry, fighting for my attention
in the form of my resources, is killing me,
it's killing you

I'm earth, when will you start to heal you
so you can begin to heal me?

I have become a dysfunctional home
and you are the variable
that enables a Lorentz transformation
that is circular in its own mistakes,
reflective
yet unchanged,

so I ask again,
desperately,
before the spacetime continuance bends
at the weight of my despair,

I am earth,
do you
truly see me?

## The Truth

If you know the way I love dreaming
then you know of my alter ego
trapped by the tempting cyclical visits of
limitless possibilities

If you know the way I remember the memories I would
like to forget then you know my id,
confused in the self-induced haze of forgetfulness
where the smoky outliers cloud the lessons
unlearned, erased, from the consciousness
of a soul struggling
hard to refract
the tracks of my infantile past

We've renounced the opportunity of happiness
by fear of criticism

At last, an honest truth

I fear as you fear

We are incarcerated by these norms that shelter expectations—
that frame the accepted portrayals of our circumstantial roles

Attained identities: a daughter, a sister, a friend then lover,
and now mother: all relations to the other
Places that set us in space
People that environ our settings
Home, school, then the workplace
People are placed in our paths
to become integrated in our experiences
Do we even truly decide the permanence of those "others"?

New responsibilities resurfacing upon my consciousness

This body has been transformed
by helical formations of learned acceptances

A new body of interpretations
to my actions or inactions
New arms that hold the heaviness of regret
New eyes to parallel the uncertainty of relation
The tone of my hurt is curt
I will nod to the given

What else is there to do?

## “Motherland”

The scream of a mother
holding her dead child
is the same,
no matter the language she utters,
no matter the flag that is waved
or the land that is claimed as the “motherland”

The frantic sobs of a father
as he recognizes the bloodied sandal of his toddler
is the same,
no matter how he prays
no matter if, when, how or where he fasts
or the land that is claimed as the “motherland”

The heaviness of the satchel
carrying the remains of his lifeless brother
is the same,
no matter the clothes that are worn,
no matter the color of his skin, eyes or toes
or the land that is claimed as the “motherland”

The sound of their loved ones voice,
the shape of their loved ones hands,
the scent of their loved ones being,
will never be the same,
because they were martyred
in the land that is claimed as the “motherland”

## Eclipse

Forward sun pushing onto the outskirts
Eclipse the clips of ice comet slips

Dismiss the mist from the crashing ocean tides that
are tugged ever so slightly by the intrusive Moon

Oceans running deep, crashing against
Stars turned numb

Black covers everything

Lighting the stroke behind the black hole
Swimming within the burping Volcano
Leading to $H_2O$
At half past 4

Gravity melts the autonomy of oneself

Colors collide into one another
Too striking to see by the naked eye
such that we hide behind the eclipse
of our own mind

Personified by dark glasses
scared of geological time

Taken from the rotation and the tilt
A phantom of stability
but not to the creator of time

I could write theories of creation
but what about foreseen destruction?
This old world being destroyed
by the temporary material deities
of planet Earth, the puppet masters
of the free (trade) world

Eclipse this reality
*Fahrenheit 451* this truth
...burn this poem to the core

The heavy iron of the written word
will not be marginalized
But centered and covered
until you see only the surface–
Basalt thoughts
Granite words
because
wind whimpers when whirled and
waters rise through the storm surge
of contradicting force

Twisted into a hurricane's grip
Mother Nature laughs at man's creations

since
steady buildings don't withhold,
houses are easily destroyed,
cures cure the symptoms not the disease

Great cities buried underneath the oceans

After divergence has paid its toll
and new ocean crust have risen to only be made old,
if not swallowed by trenches who will be recycled
then renamed,
where will we stand

but in the aftermath
of natural destruction
as our dollars bills flow freely through the air
shocked at our inability to buy our way out

## Shapes

Forming circles from triangulating
equilateral rectangles

Enduring trivial conversations
during work hours,
happy hour,
cocktail hour,
after hours,
all hours

Framing past events
with the overly decorated right angles
of wooden frames that hold together the margins
that limit the view of a life passing by
for the passerby's assumptions

Soaring through the unassuming present,
existing at a moment's time,
transcending
finite understandings

Stretching Euclidean space,
finally done talking in circles
and ready to get to the point

As I am spherically concerned,
we are constantly transformed,
yet choose to stay unchanged

No matter the dimension
or the depth
of our perception

## Quivering

A lone quivering thunder roar,
deepening its low rumble,
desperately attempting to expand its
shaking reach
across the sky that finds itself painted
by sporadic gray clouds that appear as dust—
awaken by the fresh yawn of the morning dew

I feel as a dreaded, fearful, prophetic growl that travels
in its lone circumstantial omnipresent form,
fighting its barren environmental reality

I feel in need to spread my disaster across this city

while they all still fear me, as long as I have not been
completely stripped of this fallacy of
strength and destruction

I am merely a frail wave form
that pushes air particles,
suffocated by this low-density atmosphere

A pathetic stifled cry disguised as a growl
that carries for miles across hurried allies,
whose sobs frantically race through dimly lit streets,
bouncing off depleted infrastructures that offer a
negligible shield from the torrential rain,
just to be worshiped by fear and misinterpretation
by at least one more being—one person at a time,
as they hear the thunder roar vibrate
through their shivering bodies,

I am scared that I will eventually be heard
for what I truly am

Just a quivering sigh
spreading its defeated course

## Null

I spew over pictographic designated choices,
noisily creaking from pixelated insecurities...
lately I've been trying to get to my core self...
unshelved, unsorted, undefined by a thumbs up or the "like"

traveling vibrations discharge from the speakers
of this new wave multitasking pseudo-cognitive device,
designating our leisure time into an individual undertaking of
transmitting, paradoxically, pop culture 'cross cultures through
a screen–
showcasing inputs,
transfusing collective consciousness
through neo-transcription

a mechanism whose control transcend the
pixels of equated commands ...

I miss facing a book
at night, while my daughter and son sleep in our cozy room

Instead, I turn to my screen
Lazily read about people and things
Instead of ideas
Instead of news
Instead of finding answers to questions unanswered

Reading on personal human struggles
feed the emptiness of not helping my fellow human beings
going through struggles

unheard by anyone but my mind's thought,

the care I feel for others is
fought by the lethargic tradition of
American privileged predispositions

I dislike some of the choices I've made
so I am trapped by the illusion
of an edited life

No one really knows what goes on behind the cameras,
beyond the framed shot

## A Conversation

Borges is stuck in the consumerism of salvation;
he writes of a Kafkasque hell–
a soul unconditionally trapped in the exploding cell of his
own scarlet-red perceived grandeur,
where the commercialized membrane of the protagonist's ideals
have eroded, and now, the reader is left with a post-form cell
that has penetrated continually by illusive thoughts of intellectual superiority,
brutally raped by the moral yardstick of right-wing arrogance,
where one screws themselves by criticism of what is,
while being safely cradled by the orthodoxies of the centralized church

Crusading through the
*Et Cetera* of life

What else can we count on but our own created hell?
Nevermind the control of others

Would we, ourselves, not theologians,
be stuck in our own
avenues of disguise,
disgusted at our naked bodies,
so we remain in the nude?

The connotative restrictions of being viewed
while being hidden within a shimmering red satin backdrop
under an acceptable sky-blue sky

Brecht is lost within the *End of all Games*,
stuck inside the container of a false vacuum

Even after death, even after the end

The conversation ends even though there's much to be said

## The Elusive Factories

I've never seen the factories
I have only heard of how all these
things are made

I've never seen the factories, but

I have felt the greasy screen under my fingertips,
I have held the silver metal fork to my mouth,
I have tasted the boxed food

I have only tasted
I have only touched
I have felt my belly full
I have regretted the heaviness from this food
created in factories

I've never seen the factories, but

I have smelled the scent of shit
while driving down the 110 South
near the ports

The ports that store the containers
that hold countless things:
plastic things, glass things,
ceramic things,
pharmaceutical things,
human beings...
being treated as things

A place where things come and go
From the factories to homes
From homes to packing zones
Things that make us feel more alone ...

But I've never seen the factories

And though I've never seen these elusive factories
I have seen how these factory-made things are
factoring into my valued interactions
constructing and deconstructing
my desires, morals and values

I know these factories have definite effects to my being
no matter if
I have never seen the factories

## Somewhere Else Than Here

I rather be out in the cold,
out of warm constructions,
squared buildings
and squared tones

I rather watch the trickle and the threat
of rainstorms hovering above my head

My toes turned inward
to find the shelter of my shoes' tips

My face as dry as Aphrodite's stoned lips

though I salivate when I think
of our vulnerability to the semen of the gods

Since our sustenance depends on
the variable vulnerable to the elongated equation
of causation and response

And though this mathematical equation
is not a "one-size fits all,"
it is shrunk for the purpose of patronizing ideals,
infrastructural racist appeals,
where America is the inverted number
meant to cancel out
the diffractions of *"third world"* strays
as they hang to the end,
politicians equate sanctions as being
necessary operations for equality on both ends

Both ends of one-sided benefit

The scales are tipped,
the fabric of spacetime ripped
distressed by economic "relief" packages

that supply weapons contributing to decay,
the relay race of a one-sided *his*tory, one-sided coverage,
one-sided hero, fucks given, zero

A crashed car's damage is based on
the strength of the bumpers
and the force used by the other,
both factors combine

Binary oppositions use to define
the socially constructed lies,
that are treated as truths,
implemented to marginalize

a black and white uniform,
in black and white car;
driving down the black and white housing projects of
Nickerson Gardens, criminalizing certain phenotypes

Grooming black men for incarceration
in a white system of oppression

In a world that preaches gray
but functions in black or white truths

We're all prisoners suffocated
by black and white psychological jumpers

Creating, inadvertently, racial psychoanalysis
to attempt to understand the
colored paralysis
of being black
not white
of being colored
not colorless
of being them
not us

the cognitive dissonance from
extreme binaries of existence;
living conditions that are opposing contradictions,

my region-born is not being bombarded,
my skin tone is not up for debate,
I can be left alone, be placated by
knowing that I will live another day

but others do not have assurance
so some days
I rather be somewhere else than this sphere,
full of greed, war and fear

## Low

The ceiling hangs too low,
outlined by the highest of lights,
all I can see is the shadow of emptiness,
of shallow space occupied

I need a high to keep me low...

The silence resounds within canals of waterways
Drown myself in the deafening sound of nothingness

Try to breathe through this heaviness
which environs this current shallow water quest

Sink into my subconscious will
Swim through these condensed emotions

Erode the truth that spills into the dark moles of my
written illusive temperaments

I need a low to keep me high...

With its stretched-out roads and contracted muck
The paved road comes to an end ahead

Landscape amnesia that traces the underworld into oblivion,
but some things cannot be forgotten,
as the love of Isis

Osiris was but a breath away
but Isis came a breath too late

The tears of this delay fill the Nile
with hydro-oxygenated regret

Who will the gods pray to?

I will fall to my knees
but not to pray

I pray standing or sitting,
depending on my physical state,
fate has made it so that I am rarely on my knees
(for holy reasons, at least)

Have I ever been so weak as to fall to my knees?
Have I ever fought through my shame to feel you around,
surrounding and encompassing my all?
Under this defeated soulful machinery

I need a high to keep me low,
I just need a high, goddamnit

No need to scream out to something that is a whisper away
No need to lower myself to something that will always be above me
No need to close my eyes to find you–
to close my eyes is to blind myself from you!

A maiden past, with Wiccan undertones
And yet,
I wish my heart was more faithful

Winter solstice brings forth a rebirth
And only with that end am I able to come alive
once more

## Abandoned RV Trailers

Abandoned rundown trailers down Broadway Avenue
have been there so long that the tires have gone flat
and melted to the asphalt like a surreal Dali painting

What dreams brought those RVs to the folds of LA?
What realities, like overpriced housing, kept folks
sheltered in RVs?
A choice to some, yet a necessity for others

Time dripping...

Like once upon a time runaway teenagers
on the streets,
now working them streets as used up adults.
The drug use apparent by the sagging skin
over chiseled bone,
the plastered smile over missing teeth
and the glossed gaze like
fluorescent blinding lights on
barely woken eyes
squinting,
missing the sun's orange haze settling on
our Pacific Ocean's bed
through clouds, because of draped windows,
shut blinds, locked minds... paranoid

People tripping...

What nightmares brought these folks to the
outskirts of the streets?
Kept them there.
From walking the streets to running them...
or they drop dead
since so many things are Fentanyl-laced,
placed in
a dream to euphoric allowance

Using allowances to buy pills
Overdosing in school bathrooms
as moms and dads are working overtime to hold it down
Or just mom, or just dad, or aunty, or granny...

They
Just
Drop
Dead

Feeling alone
Yet happy, so fucking happy, for just a moment,
before they fucking die,

But, fuck, can't breathe
Opioid labs, synthetic
cab... rides
Uber drivers high and drunk themselves
Riding with the windows down,
feeling the cool breeze on the
forehead and cheeks

Like a kid coming back from the AMC,
in the dark,
driving back from the big screen

Winding down
window down
feeling happy then

Happy about growing old one day...
looking forward to driving their own car
instead of being driven around
but some don't get to grow old,
and some can't afford the gas to drive their own car
or they're too intoxicated to drive
so they ride passenger side
instead

Too fucked up and need to be led

Wanting to forget the past but instead
never live in the present and die before their future

Narcan carried but given too late
And maybe you can come back from an overdose
but you can't take away the need ... the need
to fill in the hole

Life seems empty sometimes, you know
Life seems less than what life should be
So why not enhance it, enhance this,
distract me from me, into me
Take this from me
The addiction is there and will always be

Angel said it's a losing battle, that you never know
That she had overdosed a few months ago
But is back at it
Shit girl, I fucking get it
You said:
"You don't know what the fuck they put in,
I just know I want to feel a high,
want to feel alright" and yea, I get it
How could I not?

Drop Dead

Nothing changes if nothing changes
but I'm afraid of changing though

There's a homeless encampment
under the 105 overpass on Compton Avenue
approaching Imperial Avenue

There I saw a black middle-aged woman
sweeping the sidewalk next to her parked car.
She has been living out of her car for a while,
you could tell...
her car is full of things and the car
has never moved from its spot.
But, yea, she was
sweeping the sidewalk like it was her porch
and all I could do is admire the effort
because here is a woman doing
what she can to keep her space clean;
no matter how temporary,
no matter how public, or how unnecessary it may seem...

Over 20 years ago, this homeless encampment
was not fully grown like this
I walked through that underpass on the way
to my best friend's house
after school and the seedlings had begun

There were about two or three folks that would be there
posted up with beer at hand,
drinking, smoking...
friendly folks ...
always said hi when my friends and I walked by

Near that intersection there used to be a KFC that read
"we serve blacks here"...
they found it necessary to post that after the '92 riots...
ironically it was still burned down weeks
after the declaration of separation

It's like its own Favela.
Erected by the people.
Destroyed by the people

One day I saw one of the trailers down Broadway
had burned down—
Perhaps a meth lab gone ablaze
Perhaps a portable stove?
Perhaps an electrical issue?
Who fucking knows... I guess only those who lived it know
Only those close enough to the fire,
only those who feel the heat,
only those that get burned...

I am just here watching my streets,
the poverty of means painfully obvious to the naked eye,
but the wealth of worth, resilience and dignity silently rising
like the swept dust from the sidewalk to the street,
mixing with the cement but allowing the wind to gently take it
where it needs to go

## Train

Railway shaking,
a flame disrupts the solemn tone
of a citrine stone

Coal burns

Scorched in the distance,
vibrating sight,
waves of heat seen away from closure

For a time, every time, I was away spinning,
all this coal crumbling, ridges that make
black ink seem darker...

I never have been Hemmingway type of girl,
... more of a Wharton, Anzaldua, Neruda...

Trouble sleeping as I ride inside my mind

Hear my chosen music through layered metal held together
by tamed plastic,
waiting for an intersectional ring to cut the silence,
allow your shifting,
let me read some of your spilled emotions as you

stare unto me during the dead,
unsuspecting night,
both awaiting a passionate death,
come, make me come,
I will make you want...

Ironically,
love is such a tough industry, but lust is not...

but it doesn't feel the same to fuck...

Stack of dusty unread books,
unheld, unclaimed, unboxed...

I'm not really sure
of where this train is going
but I want to ride it anyway

Take it all the way down until the
coal stops burning and all that is left
is dust and ash

## Suitcase

An accumulation of moments
kept inside a beat-up powder blue suitcase
on the top shelf of
the crowded closet inside my parents' room

faded pictures held together by a yellow-stained envelope,
unhoused handwritten letters by my mother
addressed to my *abuela* in Mexico
that were never sent,
a yellow stiff dress
from when my mom graduated
as a *peinadora*
alongside
a suede dark brown drawstring bag
that holds a
loaded pistol

## The Walk

the transplanted view
that bends and folds
around my beingness

falling madly in love
with the subtle green
honey meadows that
are your eyes,
grasses swaying as we walk

all that can be trusted
is this moment's wordless
communication with nature

a spiritual asymptote
never reaching but
becoming one with the scene

## Excuse

I have an excuse to drink in the morning dew
I am a poet distressed, conveniently dressed
in a demeanor that embraces a thoughtful rue

I retreat to words for shelter in my solitude

Overused stanzas escape my cracked lips,
overworking the tercet for unearned tips,
surviving by the thrust of sluggish verbalities

Altitude above the mean so I lean into lower companionships

Grabbing the sail, almost kicking the pail by stealing the sinking ship,
steering myself from another vessel, but this one is drowning rapidly,
had to abandon ship, leave it alone, devastatingly

Overshooting the moving mark, constantly

Blackout, the white out of bright, blinding, bar lights
nonsense from fumbled verses, mumbled curses, beer flights
over loud, incoherent, scribbled noises, then... blurred nights,

Time in a half requested but forced to clockout, denied, as

language stems hang from the bend, ripped hems, he/she/them,
an audience to sneak a tiny sip of all my social slips,
...an unintended end from too much indulgence and not enough REM

## Shore

Unforgiving, jealous woman.
Crawling, turning, rowing, forming formidable mountains

spewing new earth with every undercurrent.

People drop to their knees, at her feet, to feel her more,
to indulge in her coming, to ride her...

She has been known to respond violently to sailors who do not respect her pace,
who decide to superimpose themselves over her dark blue body without consent.

It's easier to push away and let them sway, than allow an unwelcomed anchor
into her seabed.

You stand alone on the shore.

You innocently bend down to grab stones from her ocean floor;
inspiring me to come to you and innocently grab some of your attention.

I ask about the arduous hike that led you to her.

Your focus switches from her to the soles of my feet,
which are red and blistered from the hike I took on my own.

As you stare down, she stones the feet that have captivated your attention
for a fraction of a second.

I chuckle and shake my head, understanding.

I ask, "Is it ironic to covet that which leaves you feeling empty?"

You respond with, "A beauty like her, and like you, has the potential
to create and destroy."

And I stand there next to you,
realizing that you misunderstood the moment,
and believed the question was meant for you.

## Wasting Away

Am I still me, with all the passing scenes?

Temporarily moving against my own time
Vectors facing off
Shaving atoms from my skin
Wasting away
Eventually becoming an 'other'

I'm but a slither of existence

Static electricity plays with particles
Changing the slither of my atomic matter,
fluctuations around me distort
space and energy that encompass me

## An Ode to the Fibonacci Sequence

I write this for you
wonderful digits of 0, 1
1 again, 2
3, 5, and 8
13, 21, 34 and so on, and forth toward the
reflective and eternal expansive properties of life

0 the possibility of paths
1 to be redefined
1 a seedling of being
2 as the foundation
3 the birth of a pattern
5 the pentacle of power
8 the contradiction of a finite eternity

You create
gorgeous ponderous magic gaps
as you playfully approach God's favorite limit,
one ratio at a time
I am pleasantly blinded by the golden asymptotic light of your
fiery glory

The blueprints to our world,
naturally occur by your fingertip's swirl

Your ways,
entice me;
another architect at heart caught
by the sight of your "perfect plan" swag

You, oh glorious Fibonacci Code

I yearn for an intellectual discourse,
without the morning after remorse
of not having known you enough
to justify the allowance of your entering my innermost
foundational blocks of
consciousness

I'll allow your intrusive pineapple sweetness
I'll rejoice in the vibrations provided by your deep tone,
a protruding voice,
allowing
the range of your
erected constants to gently rub against the inner walls of my
ear canals, the cochlea mirroring
your moans and grunts

the outer muscles contracted, perked, excited to hear more

You send impulses to my
slightly
swollen
pink
brain

Climaxing when I willingly accept the repeated
input of your aptitude;
the elongated numerical sequence
you utilize to reproduce is
breathtaking

I write this to you
the sequence that echoes God's voice

You are the dispersed seeds of the forbidden fruit
You are the leaf arrangement of the burning bush
You are the spatial range of the nautilus

You are the canny lover that strategically bestows
knowledge from the forbidden tree

## Acknowledgments

In my attempt to chronicle an imperfect love life through poetry I came to uncover that poetry itself is the muse. The way it inspires, repurposes and shapes experiences into structured consciousness; holding within it the spirit of unconscious discovery. This book is born from the love of writing, the love of human connection, and the potential both hold in shifting perspective and emotion.

I want to thank my dad for subconsciously instilling in me the importance of the written word through our daily walks to Jay's Liquor Store to buy La Opinión. Our morning expeditions showed me that certain things in life are worth the walk. I would like to thank my ma-motz for teaching me to love deeply but to always leave enough for myself, if ever I need to walk away. Through her actions I found that love, resilience and acceptance take on different forms, and not all are self-sacrificing. She embodied strength and kindness in all her interactions with people, places and circumstances, so I carry myself in that same spirit. I am profoundly grateful to my children for blessing me with the constant reminder of my humanity; the range of emotions I experience day-to-day with them is vast, messy but beautiful. Finally, I am eternally indebted to the strangers who became friends, mentors, lovers and lessons learned. Without you, there is no Ion Love.

## Previously Published Material

Previously published online on Poets for Science.

Gaytán, Julia C. "An Ode to the Fibonacci Sequence." December 25, 2025

Gaytán, Julia C. "Ion Love." December 25, 2025

Previously Published in Chicana/Latina Studies by Mujeres Activas En Letras y Cambio Social

Gaytán, Julia C. "Imported Sand" and "Suitcase." Chicana/Latina Studies, vol. 24, no. 2, MALCS Spring/Summer 2025, pp. 142–47.

Previously Published in The Waiting Room by Nervous Ghost Press

Gaytán, Julia C. "Impossible Encounter with the Other Side of Our Truth" The Waiting Room, vol. 1, Nervous Ghost Press, 2023, pp. 168-69.

Gaytán, Julia C. "ACT III: SCENE 2" The Waiting Room, vol. 1, Nervous Ghost Press, 2023, pp. 200-01.

## About the Author

Julia C Gaytán is a Xicana poet and math geek who spends her weekdays teaching and her weeknights exploring the city with friends, writing poetry, reading and watching anime. Julia grew up in Nickerson Gardens, a housing project of Watts, and attended inner-city k-12 public schools until attending UCLA as an undergraduate then graduate student. She holds majors in Chicana/o Studies and History with minors in African American Studies and Political Science. It is in her studies that she fully began to understand her community within a larger world context and develop new frameworks to explore identities and realities outside of her lived experience while simultaneously embracing her identity as a Xicana poet. Her work is published in MALCS Chicana/Latina Studies Journal and The Waiting Room, an anthology by Nervous Ghost Press. Currently, she organizes microcinema screenings for Espiral Collective and participates in various speaking engagements around Southern California.

## About the Publisher

Riot of Roses Publishing House is a radical feminist, award-winning press that was founded in 2021 to amplify the stories of historically silenced voices and narratives.

Xicana owned. Mujerista focused. For the people.

We publish books that heal and liberate.

Read our rebellion.

Find & follow us @riotofrosespublishing
Visit us at www.riotofrosespublishinghouse.com

Riot of Roses Publishing House offers bulk quantity discounts to educational institutions, businesses, and community groups. For more information, reach out to us at riotofrosesllc@gmail.com

RIOT OF ROSES
PUBLISHING HOUSE
SEJATNGA
UNCEDED TONGVA TERRITORY
SOUTH WHITTIER, CALIFORNIA

www.ingramcontent.com/pod-product-compliance
Lightning Source LLC
Chambersburg PA
CBHW031351230926
51848CB00002B/479
*9781961717503*